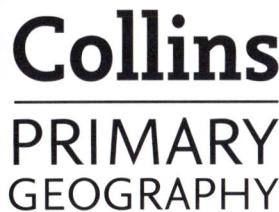

World around me

Pupil Book 1

Planet Earth	Air	4
	Water	6
	Soil	8
	Rock	10
Weather	Sun	12
	Rain	14
	Cold	16
	Wind	18
Local area	Home	20
	Shops	22
	School	24
	Park	26
Food and farming	Vegetables	28
	Fruit	30
	Bread	32
	Milk	34
Habitats	Woods	36
	Grass	38
	Sand	40
	Ponds	42
Journeys	Walking	44
	Wheels	46
	Cars and lorries	48
	Buses and trains	50
Mapwork skills	Colour	52
	Near and far	54
	Low and high	56
	Signs	58
Maps and plans	Maps and plans	60
	Our world	62

Stephen Scoffham | Colin Bridge

World around me

Geography is about the world and where we live.

It is about our homes and school, and the places close by.

Geography helps us to know about the countryside, the weather and our food.

It shows how we share the world with plants and animals.

Geography is about maps. They are pictures of the world.

Become a geographer.

Know what is special about the places near your home.

Pip and Tom are going to help you learn more about geography.

Pip   Tom

Air

Air is all around us

Read and talk about the story.

Pip blows some bubbles full of air.
The big one pops – bird has a scare!

Yes or no?

1. The girl is blowing air.
2. The bubbles are full of water.
3. The bird flies in the air.

Talking

What three things can you do by blowing air?

Things float in the air

Choose the best word.

❶ The balloon is flying in the …

[air]   [water] .

❷ The cloud is in the …

[sea]   [air] .

❸ The butterfly has …

[air]   [sun]

under its wings.

Look and find
What is moving in the air? How many birds can you see?

Water

The world is full of water

Read and talk about the story.

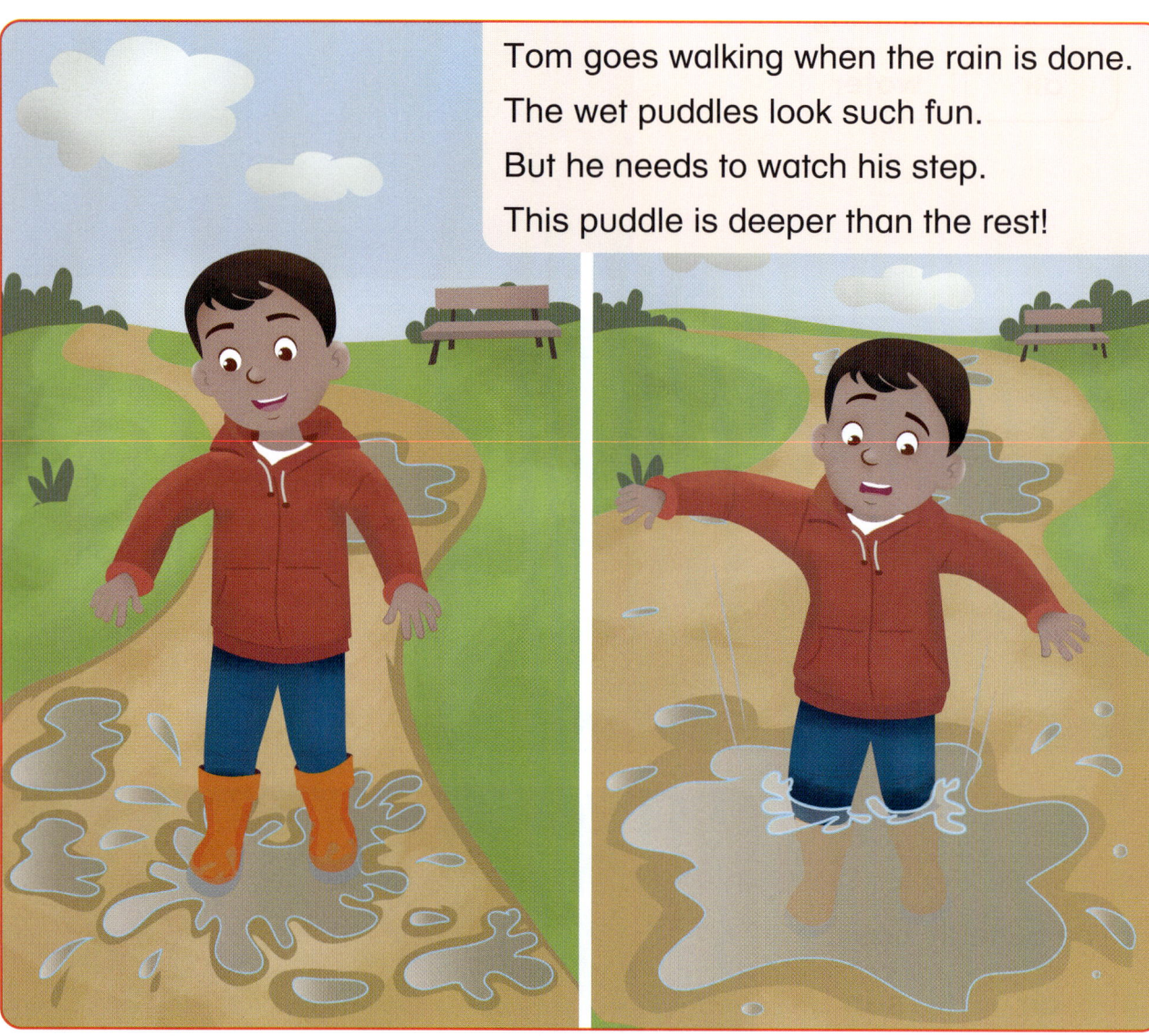

Tom goes walking when the rain is done.
The wet puddles look such fun.
But he needs to watch his step.
This puddle is deeper than the rest!

Choose the best word.

puddles   taps   clouds

1. Water makes …
2. Water falls from …
3. Water runs out of …

Talking
What do you like doing with water?

We can see water

Match the words to the pictures.

| river | pond | waterfall |

We can use water

True or false?

① Plants need water to grow.

② Food is cooked in water.

③ We wash in milk.

Look and find
How many taps can you find in your school?

Soil

The ground is made of soil

Read and talk about the story.

Worm makes a burrow.
Mole makes a hill.
Mouse takes a tumble.
Worm looks on still.

Choose the best word.

1. The mole has made a … mill   hill   till .

2. The trees grow in the … sky   grass   soil .

3. The worm is in the … soil   water   sky .

Talking
What things in the story and picture need soil?

Living things need soil

Choose the best word.

| field | garden | wood |

1 Flowers grow in soil.
The soil is in the ...

2 Cabbages grow in soil.
The soil is in the ...

3 Birds find food in soil.
The soil is in the ...

Look and find
Can you find soil around your school? Is anything in it?

Rock

Rock makes solid ground

Read and talk about the story.

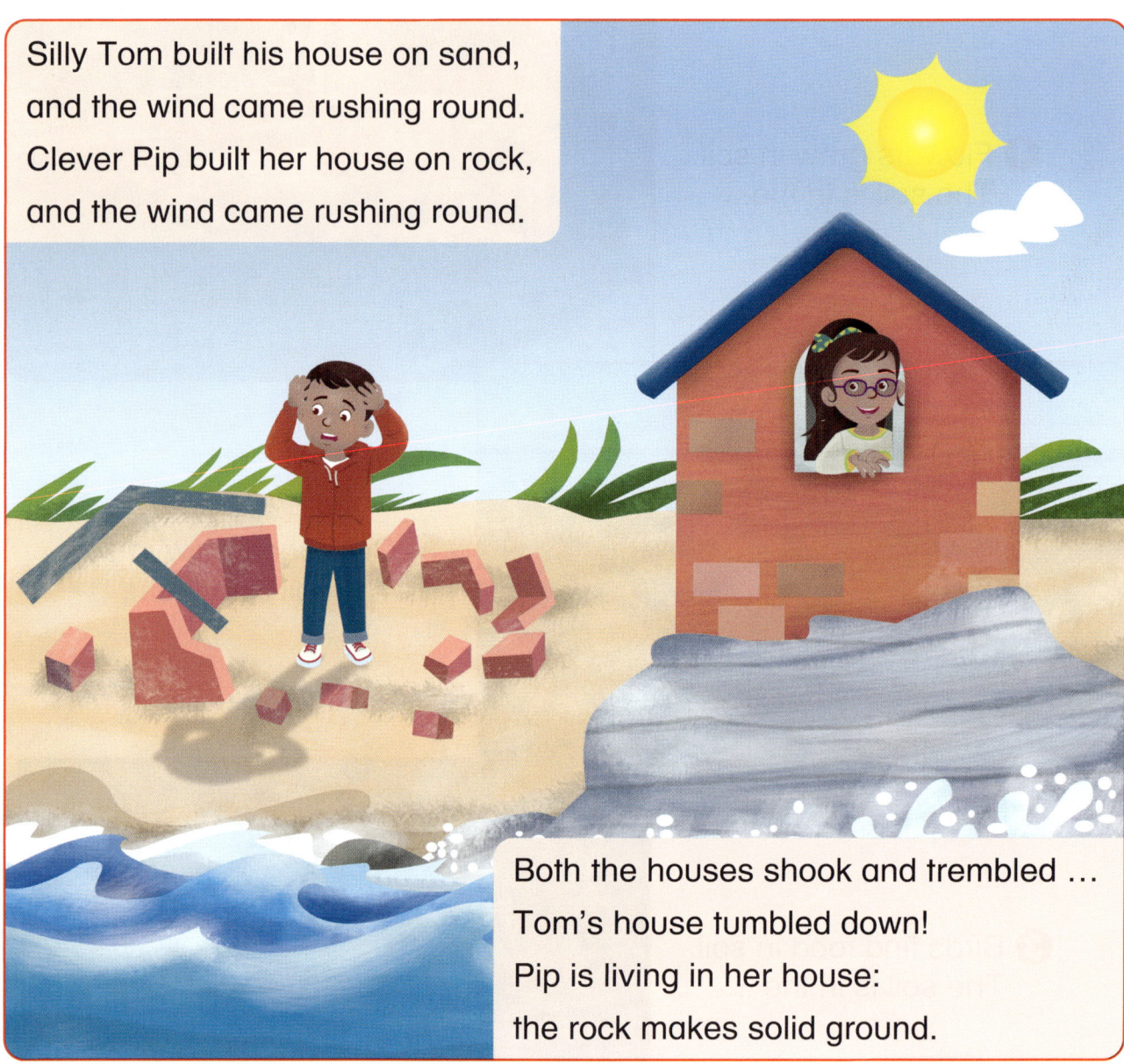

Silly Tom built his house on sand, and the wind came rushing round. Clever Pip built her house on rock, and the wind came rushing round.

Both the houses shook and trembled … Tom's house tumbled down! Pip is living in her house: the rock makes solid ground.

Choose the best word.

| soft | hard | rock | sand |

1. Tom built his house on …
2. Sand is …
3. Pip built her house on …
4. Rock is very …

Talking
How are houses fixed into the ground?

Rocks are different shapes and sizes

Match the words to the pictures.

pebbles   boulders   stones

1 small and round

2 hand-sized

3 very big

True or false?
1. I can throw a pebble.
2. I can see stones in gardens.
3. I can carry a boulder.

Look and find
Is rock used anywhere at your school?

Sun

The sun is very hot

Read and talk about the story.

Out comes the sun.
It's very hot.
I want this ice-cream
quite a lot!

Look! Now it's just
an empty cone.
It's still so hot.
I shall go home.

Choose the best word.

1. The sun is … [shining] [hiding].
2. Ice-cream is … [hot] [cold].
3. The sun makes things … [melt] [freeze].

Talking
What can you do on a sunny day?

The sun is a ball of fire in space

Look at the picture.

Choose the best word.

| sun | Earth | moon |

1. I can see the hot …
2. We live on the …
3. I can see the little …

We need the sun

Find these things in the pictures.

Light from the sun helps us …

| read | walk | cycle |

Warmth from the sun helps …

| flowers to grow | bees to fly |

| washing to dry |

Look and find

Does the sun shine into your classroom in the morning or the afternoon? Where are the darkest parts of the room?

Rain

Rain falls from the sky

Read and talk about the story.

I am in my coat and boots.
The clouds above float round.
It's fun to play on a wet day,
with water all around.

Oh no! The sun has come back out.
It's far too hot again.
The only water left to see
is puddles made of rain.

Choose the best word.

| clouds | puddles | wet |

1. When it rains the weather is …
2. Rain falls from the …
3. Water from rain makes …

Talking
What clothes do you wear on a wet day?

Wet weather changes

Choose the best words.

drizzle

1 Drizzle is when it rains …

| a lot | a little |

a shower

2 A shower is when the rain lasts …

| a short time | all day |

a storm

3 A storm is when the rain is …

| gentle | very heavy |

Look and find
Where does rain land on your school? Where does the rainwater go?

Cold

Cold weather changes what we do

Read and talk about the story.

I like white snow.
It's cold and crisp.

Here comes a snowball.
Make it miss!

Choose the best word.

1. We have snow when the weather is [hot] [cold].

2. Snow is frozen [rain] [stones].

3. Snow falls from the [clouds] [sun].

Talking
What words might you use to talk about a cold weather day?

Cold weather freezes rain

Yes or no?

snow

❶ The snow is in the sky.

ice

❷ The ice is on a puddle.

icicles

❸ The icicles are on the ground.

Look and find
What keeps us warm in school?

Wind

Wind changes the weather

Read and talk about the story.

In the United Kingdom, the wind changes the weather.

The compass tells us the wind direction.

The North Wind is cold.
The East Wind is dry.
The South Wind is hot.
West Wind blows the rain by.

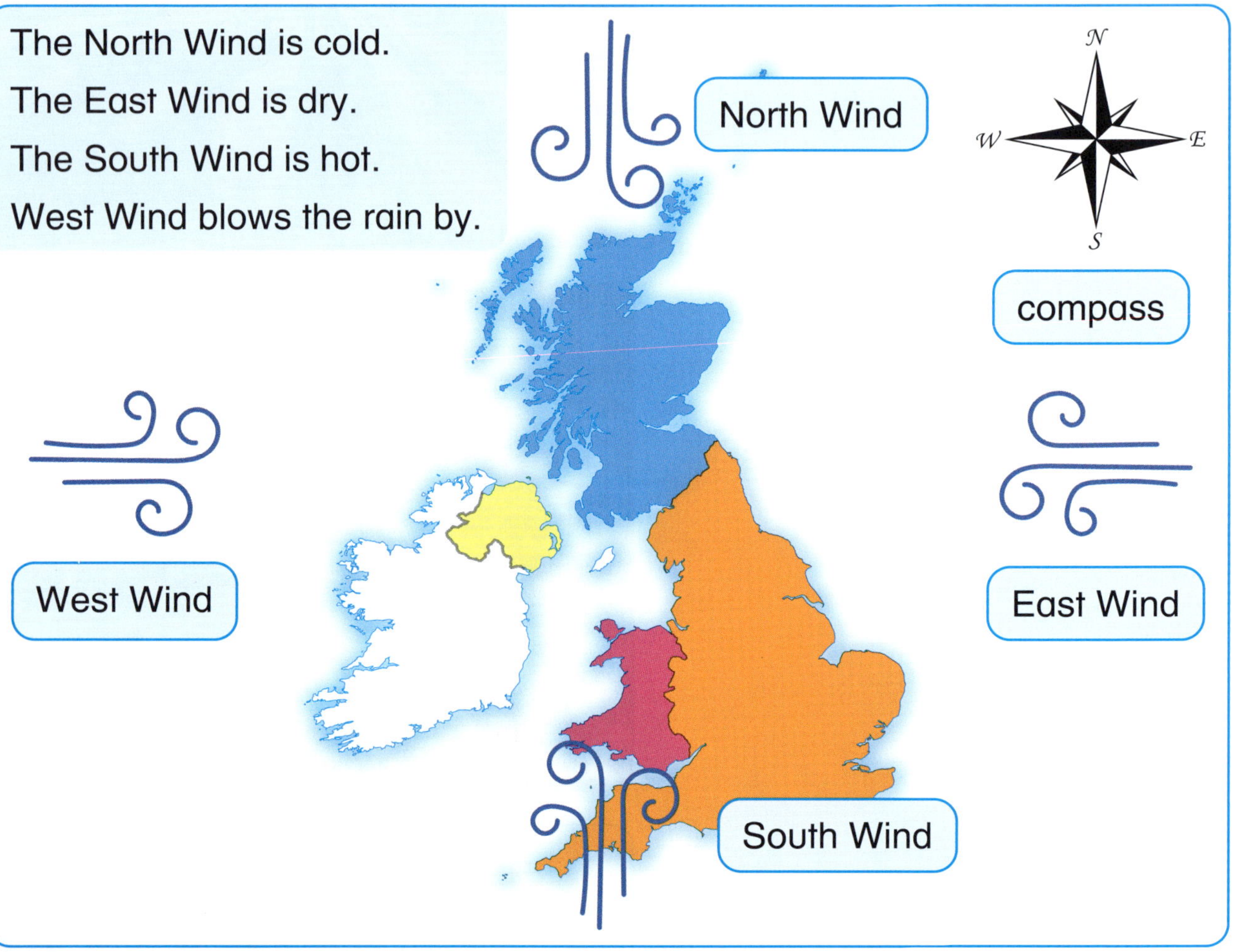

North Wind

compass

West Wind

East Wind

South Wind

Choose the best word.

| dry | cold | hot | rainy |

1. The North Wind makes the weather …
2. The West Wind makes the weather …
3. The South Wind makes the weather …
4. The East Wind makes the weather …

Talking
Where are North, South, East and West in your classroom?

The wind moves and blows

Spot the difference.
There are four to find.

Wind is not always the same

still

no wind

breeze

little wind

What is the wind like today?

gale

strong wind

Look and find
Which days are windy this week?

Home

Home has the things we need

Read and talk about the story.

Tom has left his toys lying on the floor.
Now he wants to play with them.
Help him find them all once more.

Choose the best word.

| bedroom | bathroom | kitchen | living room |

1. The car is in the …
2. The bat is in the …
3. The ball is in the …
4. The duck is in the …

Talking
Why are there different rooms in a house?

We use places in the home

How do we use rooms?
Choose the best word.

1. Cooking is in the …
   bedroom   **kitchen**

2. Washing is in the …
   bathroom   **garden**

3. Sleeping is in the …
   kitchen   **bedroom**

We can do jobs at home

Yes or no?

1. I can tidy up my toys.
2. I can set the table.
3. I can put rubbish in the bin.
4. I can wash the dishes.

Look and find
What three things can you do at home and at school? What things can you only do at home?

Shops

Shops sell the things we need

Read and talk about the story.

Pip is going shopping now. She has to make a choice.

She looks at all the many things, but everything looks nice.

Will she buy a pack of sweets, or maybe an apple pie?

She takes so long to make a choice. "Too late!" the shopkeeper cries.

True or false?

1. Shops sell things.
2. All shops sell food.
3. Shops give things away.

Talking

What is your favourite shop?

Shops are not the same

Choose the best word.

1. This shop sells …
   bread   chairs .

2. This shop sells …
   carrots   toys .

3. This shop sells …
   everything   kangaroos .

We need things to go shopping

Which of these things do you need to go shopping?

 money

 football

 a list

 a bag

 a chair

Look and find
Which shop do you go to most? What does it sell?

School

Schools help us to learn

Read and talk about the story.

Pip does not want to go to school today.

She would like to stay at home and play with toys all day.

But Friday lunch is fish and chips.

And Pip begins to lick her lips.

Yes or no?

1. At school we learn to read.
2. At school we bring our toys.
3. At school we paint and draw.
4. At school we go to sleep.

Talking
What part of the classroom do you like best?

There are places and spaces in a school

Find the places in the picture.

| playground | classroom | hall | cloakroom |

Where do children do these things in a school?

1. Play
2. Eat
3. Paint
4. Put away coats and bags

Look and find
What two things do you like doing in your school?

Park

Parks are fun and healthy

Read and talk about the story.

This place is just for children, but Grandad came in too.

Grandad is in the tunnel. Now he is stuck like glue!

What is the right order?

1. Grandad is stuck.
2. Tom is in the tunnel.
3. Grandad wants to have a go.

Talking

What do you like best in a park?

We can do lots of things in a park

Choose the best word.

| swinging | spinning | climbing |

Space number 1 is good for …

Space number 2 is good for …

Space number 3 is good for …

Which space number is good for little children?

Look and find

What can you do in different spaces in your school playground?

27

Vegetables

Vegetables help our bodies to grow

Read and talk about the story.

Pip likes to eat her vegetables.
She fills a great big sack.

Now she tries to take them home.
They are too heavy for her back.

Choose the best word.

| green | round | orange | oval |

① Cabbages are …
② Carrots are …
③ Onions are …
④ Potatoes are …

Talking
What vegetables do you like to eat?

Farmers grow vegetables

Find the things in the picture.

- farmer
- onions
- carrots

Look at the pictures and find how many there are.

How many potatoes?

How many cabbages?

How many onions?

How many carrots?

True or false?

All vegetables grow under the soil.

Look and find

Which of these vegetables have you seen growing: potatoes, carrots, onions, cabbages?

Fruit

Fruit is sweet and healthy

Read and talk about the story.

My little apple tree is bare. It is the winter time.

In spring the leaves begin to come. Still the apples are not mine.

In summer there are flowers to see. The tree is very pretty.

At last the autumn brings the fruit, and I shall go to pick it.

Choose the best word.

| spring | summer | autumn | winter |

❶ The tree is bare in …
❷ The leaves come in …
❸ There are flowers on the tree in …
❹ You can pick the fruit in …

Talking
What are the different parts of an apple?

There are different kinds of fruit

Match the words to the pictures.

plum   cherry   pear   strawberry

True or false?

1. Fruit grows under the ground.
2. Fruit grows on trees.

Look and find
Can you find any fruit in your school today?

Bread

Bread is made from seeds

Read and talk about the story.

Tom plants seeds to turn to flour and bake it into bread.
He asks two friends to help him, but they say, "No!" instead.

The mill grinds seeds to flour.
The baker bakes a tasty loaf.
His friends now want to have some, but Tom says, "No!" to both.

Choose the best answer.

| planting seeds | baking bread | grinding seeds |

1. What is Tom doing?
2. What happens at the bakery?
3. What happens at the mill?

Talking
What are the names of different sorts of bread?

We can make bread

Are all loaves the same?

Look at the photos. **What is the right order?**

Look and find
Flour can be baked into a loaf, rolls, bread sticks, wraps, biscuits and cake. Which do you like the best?

Milk

Milk is drink and food

Read and talk about the story.

The children are on holiday.
They go walking by a farm.
Is that a snake? They start to run,
so they will not come to harm.

But they are being silly.
That snake is just a tail!
The cows are here for milking time.
They come each day and never fail.

Tell the story of the children's walk. Use these words in your story:

- farm
- milking
- tail
- snake

Milk comes from cows

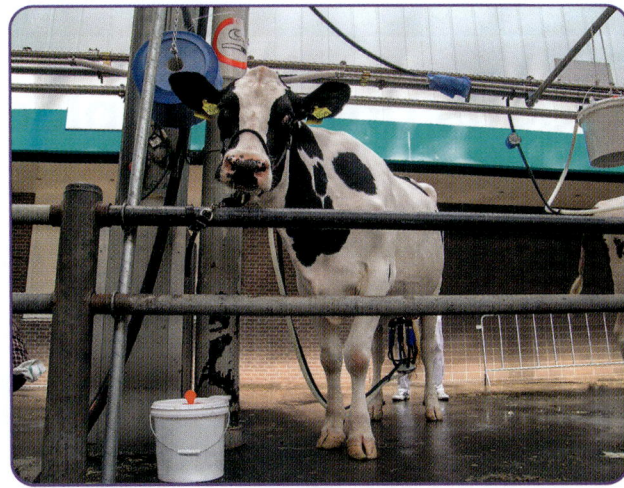

True or false?

1. Cows eat bread.
2. Cows make milk.

Other food is made from milk

Match the words to the pictures.

- butter
- cheese
- yoghurt

Talking
Ask your friends: Who likes cheese? Who likes yoghurt? Which wins?

Look and find
Does milk come to your school? How does it get here?

Woods

Woods are shady places

Read and talk about the story.

Tom is out on holiday.
He walks the lonely wood.

He would like some friends along,
if only that he could.

Tom trips over on a root,
and falls down to the ground.

Now there are lots of animals
coming all around.

Find the woodland animals.

rabbit   fox   badger   squirrel   woodpecker

Find the woodland plants.

trees   bluebells   grass   toadstool

Talking
Why do so many animals like to live in woods?

The wood is a home

Choose the best answer.

| nuts | badger | fox |

❶ In the wood there are holes to live in.

What animal lives here?

❷ In the wood there is food.

What is the squirrel eating?

❸ In the wood there is water.

What animal is drinking?

Yes or no?

❶ The wood is a home for plants and animals.
❷ This home is called a habitat.
❸ Animals find food and water in a wood.
❹ It is good to cut down trees.

Look and find
Count the trees you can see around your school.

Grass

Grass makes open spaces

Read and talk about the story.

Tom sits down upon the grass to have something to eat.

The grass is soft and all is quiet. He starts to fall asleep.

Now he is hungry. He sits up and reaches for some food.

Oh no! Ants run off with the bread. Snails eat the lettuce too.

Choose the best word.

1. Grass is [blue] [green].
2. Grass is [soft] [hard].
3. Snails eat [leaves] [stones].
4. Ants take food into their [cars] [nests].
5. Under the grass are [worms] [birds].

Talking

Why do flowers grow in grassy places?

Grass is a home

Look at the picture. **Find how many there are.**

1. How many butterflies?
2. How many worms?
3. How many slugs?
4. How many ants?
5. How many snails?

butterfly

worm

slug

ant

snail

True or false?

1. Nothing lives in grass.
2. Grassy homes are called habitats.

Look and find
Where does grass grow in your school? Look high and low.

Sand

The seashore is a changing place

Read and talk about the story.

Tom is down by sand and sea.
It is a sunny day.
"First I will finish my ice-cream,
then I will go and play."

Watch out! The beach is busy.
Many creatures live nearby.
Crab nips at Tom's big toe.
Gull takes the ice-cream up sky high!

Choose the best word.

| grains | sand | rock |

1. The beach is made of …
2. The sand is made of tiny …
3. The grains are little bits of …

Talking

Seashore creatures have to hide away. How does the sea change every day?

A sandy shore is a home

Look at the picture. **Count how many there are.**

1. How many crabs?
2. How many gulls?
3. How many starfish?
4. How many lugworms?
5. How many limpets?
6. How many winkles?
7. How many razor clams?

crab

gull

starfish

lugworm

limpet

winkle

razor clam

Look and find
Find a seaside story. Draw a picture of a plant or creature from the story.

Ponds

Ponds are a water habitat

Read and talk about the story.

Tom's holiday is over.
He is feeling rather glum.

He walks on past the duck pond,
and he is nearly home.

He crumples up a plastic bag,
and throws it in the pond.

He stops and thinks, then picks it up.
"To spoil a home is wrong!"

Choose the best word.

| water | fish | home | habitat | food | ducks |

1. A pond is made of …
2. In the water live …
3. This home is called a …
4. Creatures make ponds their …
5. Swimming on top are …
6. At the edge frogs find …

Talking
Why is it wrong to throw rubbish into a pond?

A pond is a home

Look at the picture. **Find these creatures.**

| frog | dragonfly | fish | snail | tadpole | duck | beetle |

Look and find

Where could you make a little pond at your school?

Walking

Walking is for short journeys

Read and talk about the story.

"Get up! Get up!" says Pip to Tom. "Granny is not well today."

"We must take this cake for her to eat. Walking is the quickest way."

True or false?
1. Walking helps me go where I want to go.
2. Walking is good exercise.
3. Walking is free.
4. I see things when I walk.
5. I can walk with my friends.

Talking

Is there a place you like walking to?

We can walk to places

Look at the picture. Pip can walk to nine places. **Find the places.**

playground | duck pond | café | school | shop | library

museum | friend's house | ice-cream van

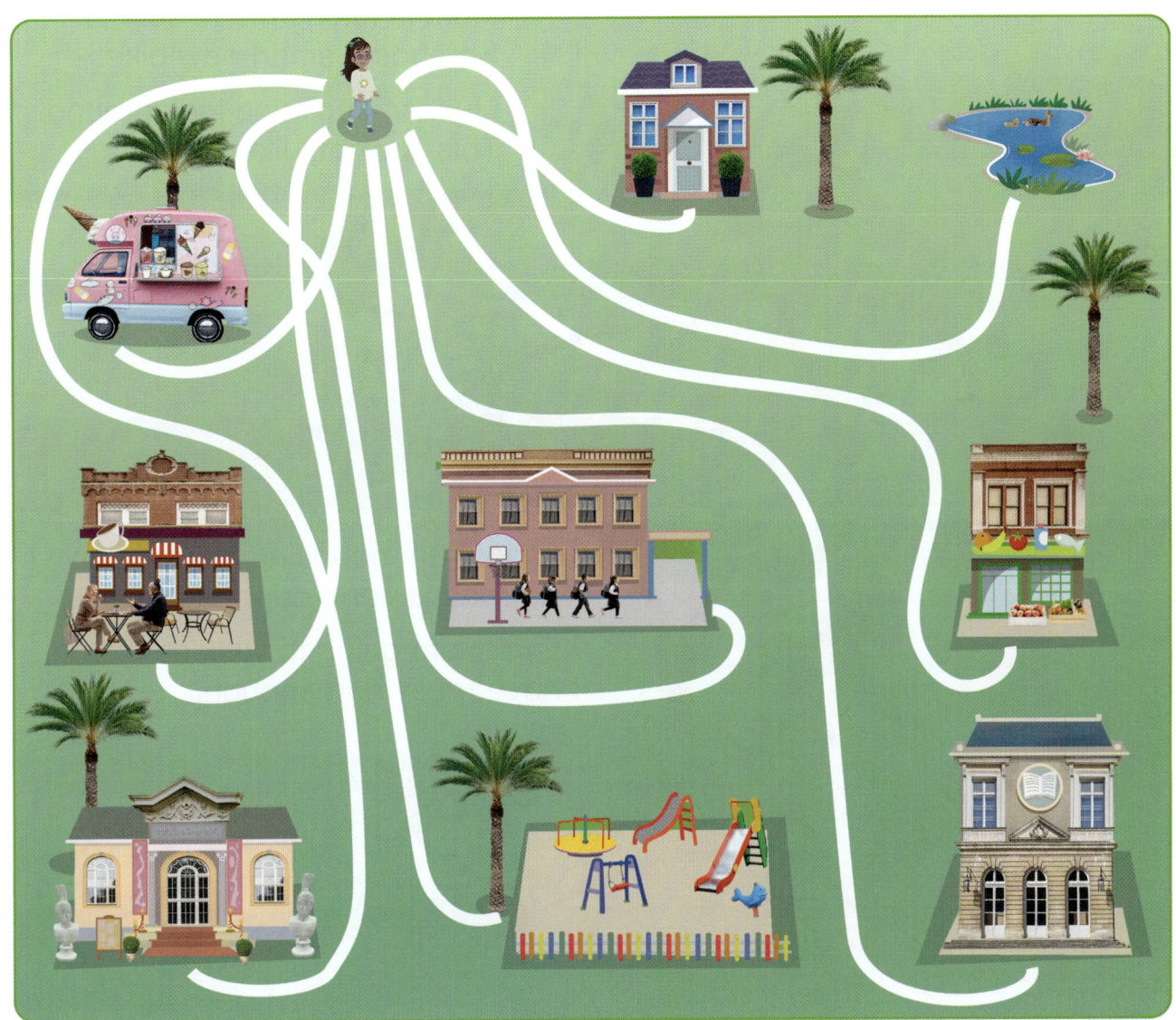

Look and find

What three places in your school can you walk to from your classroom?

Wheels

Wheels make moving easy

Read and talk about the story.

Mum is taking me to school. She has a smart new bike.

I just sit back and look around. This is what I like.

Oh no! I am bouncing up and down. The road is very rough.

I think I am going to walk next time. I have really had enough.

True or false?
1. Wheels are square.
2. Wheels make moving easy.
3. Bikes have four wheels.
4. Bikes go by themselves.

Talking
What is your favourite thing with wheels?

Wheels help us travel

Look at the picture.

What things with wheels can you see?

Which of these can make long journeys?

Which are for fun?

bicycle

skateboard

baby buggy

scooter

children's bike

Look and find

Ask your class: How many children have bikes? How many have scooters?

Cars and lorries

Cars and lorries carry loads

Read and talk about the story.

I have nearly reached the play park. It's just across the road.

I look both ways and try to cross, just as I have been told.

Oh dear! The cars and lorries have filled the road right up.

Will I ever get across? This really is bad luck.

Yes or no?
1. Cars have five wheels.
2. Cars are for people.
3. Lorries make long journeys.
4. Lorries carry the things we need.

Talking
Would you like to drive when you grow up?

Cars and lorries make journeys

Find these cars, vans and lorries in the pictures.

| car | food lorry | taxi | ambulance |

| tanker | post van | dustbin lorry | ice-cream van |

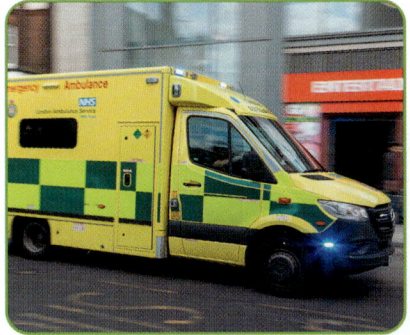

Look at the pictures.

Which ones carry people?

Which ones carry things?

Look and find
What do lorries and vans bring to schools?

49

Buses and trains

Buses and trains make long journeys

Read and talk about the story.

Dad says, "It's time for holiday.
We shall go far away."
"No! No!" I say. "It is not fair.
I cannot walk all day!"

Dad says, "Tom, do not worry.
The journey is too long.
We are going to the station,
and the train will come along."

Yes or no?

1. Trains and buses carry many people.
2. Buses go on rails.
3. Trains are very speedy.
4. Buses have many stops for people to get on.

Talking
Why do buses have numbers on the front?

Buses and trains use stations

True or false?
1. Buses are all the same.
2. Buses make journeys.
3. Trains stop at stations.
4. Trains go on water.

Look and find
Make a class exhibition of toy buses and trains.

Colour

Maps use colour

Read and talk about the story.

Pip is looking at her map.
She wonders what the colours mean.
She looks down from a grassy hill,
and this is what she sees.

Yes or no?
1. Green is for the countryside.
2. Blue is for the road.
3. Grey is for the town.
4. Yellow is for the sky.

Talking
Why do maps use colours?

Colours send messages

Choose the best word.

green   red   yellow

❶ ... means stop and think.

❷ ... means be careful.

❸ ... means you can do it.

Look and find
Find colour messages in red, yellow and green around your school. Which colour is used most?

Near and far

Places can be near or far

Read and talk about the story.

A squirrel sits up in a tree.
It's tired of being there.
It can see a better tree.
It seems to be quite near.

That tree looks near, so off it goes.
It jumps into the air.
Oh dear! The squirrel's falling to the ground!
It really was too far.

Choose the best words.

1. The squirrel is sitting in a [chair] [tree].
2. It can see another tree [near] [far off].
3. It jumps across and [gets there] [falls down].
4. The tree is too [near] [far].

Talking
What part of your school is far away from you now?

Maps show places near and far

Look at the picture. The bus stops in lots of places.

| shop | school | park | café | farm house |

Where does the bus stop that is near?

Where does the bus stop that is far away?

Look and find
Look out of the window or across the playground. Which things are far away?

Low and high

Things can be low or high

Read and talk about the story.

Pip has a brand new kite.
She wants to make it go.
There is no wind to take it up.
This place is very low.

Now this is better! Up it goes.
It's flying in the sky.
Up on this hill the wind is strong.
The kite is flying high.

Choose the best word.

1. Pip has a new [kite] [clock].

2. It won't fly because there is no [rain] [wind].

3. Pip goes up a [hill] [tower].

4. The hill is windy because it is [low] [high].

Talking
What three things are high up in your classroom?

Maps show places low and high

Look at the picture.

Which things are up high?

Which things are down low?

| seat | plane | bike | windmill | bin | lighthouse |

True or false?

1. Planes fly from low to high.
2. Water runs from low to high.

Look and find

Go outside and look at your school. What parts are high up? What things are low down?

Signs

Signs tell us what to do

Read and talk about the story.

Tom is at the seaside but he is feeling a bit sad. The signs all seem to tell him that so many things are bad!

This is looking better! He sees some friendly signs. They show some things he can do. He will have a jolly time.

Yes or no?

When Tom is at the beach he can …

1. ride a bike.
2. eat an ice-cream.
3. camp in a tent.
4. go swimming.
5. fly a kite.

Talking
What signs are important for children to know?

Signs tell us messages

Match the words to the pictures.

sad   happy   angry   thinking

Match the words with the signs.

castle   sports   holiday

Look and find
What part of your playground do you like the best? Make a blue sign to show what you can do there.

Maps and plans

Street maps and plans show where things are

Read and talk about the story.

Tom has a plan to see where his new chair can go.

Pip has found a street map to ride home up the roads.

Choose the best word.

| room | roads | above |

1. Tom's plan shows his …
2. Pip's street map shows the …
3. Maps and plans show places from …

Talking

What can a map tell you?

Maps and globes show where things are

Read and talk about the story.

Tom sits high up in a tree and looks down from above.

This special map is like a ball: the globe shows Planet Earth.

Pip is off on holiday to find fun things to do.

This map shows all the things there are, like castles and a zoo.

Choose the best words.

| globe | zoo | ball | castle | railway | ship |

1. Tom's map is called a …
2. A globe is shaped like a …
3. On Pip's map, there is a …

Look and find

Look at a globe. Find your country.

Our world

We must care for the world

Read and talk about the story.

Our friends now have to say "Goodbye!"
They try to do their best.
They share the world with you and me, with creatures and with plants.
So shall we help them keep it well? The answer must be "Yes!"

> **Talking**
> What can we do to help the world?

This is the United Kingdom

Look at and talk about the picture.

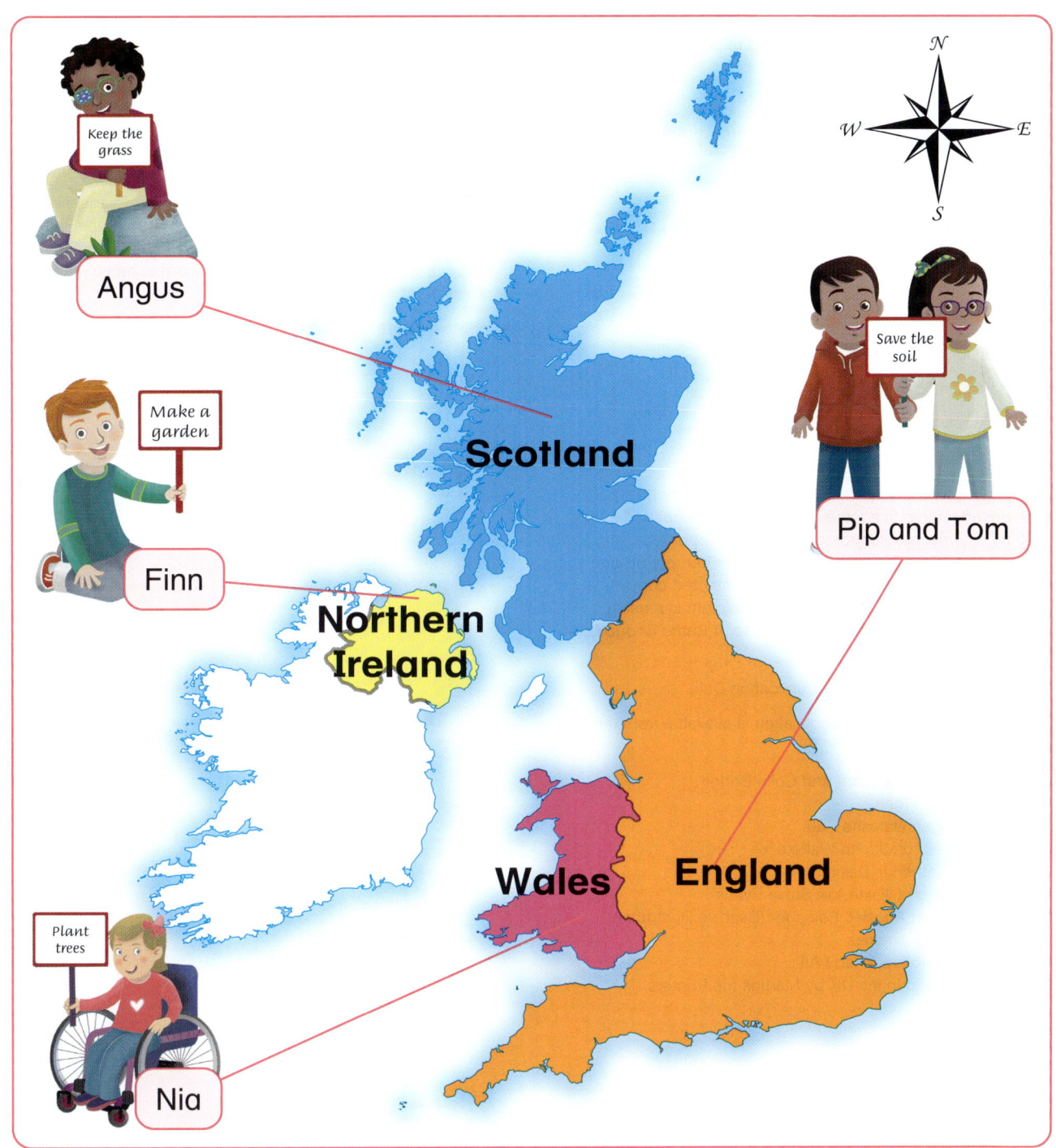

Look at the picture.

Where do Nia, Finn, Angus, and Pip and Tom live in the United Kingdom? England, Scotland, Northern Ireland or Wales?

Look and find

Look at your school's address. What does each line mean?

William Collins' dream of knowledge for all began with the publication of his first book in 1819.

A self-educated mill worker, he not only enriched millions of lives, but also founded a flourishing publishing house. Today, staying true to this spirit, Collins books are packed with inspiration, innovation and practical expertise.
They place you at the centre of a world of possibility and give you exactly what you need to explore it.

Published by Collins
An imprint of HarperCollins*Publishers*
The News Building, 1 London Bridge Street, London, SE1 9GF, UK

HarperCollins*Publishers*
Macken House, 39/40 Mayor Street Upper, Dublin 1, D01 C9W8, Ireland

Browse the complete Collins catalogue at
collins.co.uk

© HarperCollins*Publishers* Limited 2025

Maps © Collins Bartholomew 2025

10 9 8 7 6 5 4 3 2 1

ISBN 978-0-00-872828-1

All rights reserved. No part of this publication may be reproduced, stored in a retrieval system, or transmitted in any form by any means, electronic, mechanical, photocopying, recording or otherwise, without the prior written permission of the Publisher or a licence permitting restricted copying in the United Kingdom issued by the Copyright Licensing Agency Ltd, 5th Floor, Shackleton House, 4 Battle Bridge Lane, London SE1 2HX.

British Library Cataloguing-in-Publication Data

A catalogue record for this publication is available from the British Library.

Authors: Stephen Scoffham and Colin Bridge
Publisher: Laura White
Product manager: Natasha Paul
Development editor: Judith Walters
Proofreader: Catherine Dakin
Cover designer and illustrator: Steve Evans
Internal illustrator: Ángeles Peinador (Beehive Illustration)
Typesetter: David Jimenez
Production controller: Alhady Ali
Printed and bound in the UK by Martins the Printers

This book is produced from independently certified FSC™ paper to ensure responsible forest management.

For more information visit: www.harpercollins.co.uk/green
collins.co.uk/sustainability

Acknowledgements

The publishers gratefully acknowledge the permission granted to reproduce the copyright material in this book. Every effort has been made to trace copyright holders and to obtain their permission for the use of copyright material. The publishers will gladly receive any information enabling them to rectify any error or omission at the first opportunity.

All photos: Shutterstock